Henry Cavendish

Experiments on air

Papers published in the Philosophical transactions

Henry Cavendish

Experiments on air
Papers published in the Philosophical transactions

ISBN/EAN: 9783742827227

Manufactured in Europe, USA, Canada, Australia, Japa

Cover: Foto ©ninafisch / pixelio.de

Manufactured and distributed by brebook publishing software
(www.brebook.com)

Henry Cavendish

Experiments on air

EXPERIMENTS ON AIR.

PAPERS PUBLISHED

IN THE

PHILOSOPHICAL TRANSACTIONS

BY

THE HON. HENRY CAVENDISH, F.R.S.

(1784-1785.)

EDINBURGH:

WILLIAM F. CLAY, 18 TEVIOT PLACE.

LONDON:

SIMPKIN, MARSHALL, HAMILTON, KENT & CO. LTD.

1893.

PREFACE.

THE two papers contained in this reprint both appeared in the Philosophical Transactions under the title "Experiments on Air." The first paper, which was published in 1784, contains Cavendish's account of his investigations into the composition of water. This paper is of great historical interest and importance in connection with "the Water Controversy,"as the discussion concerning the first discovery of the composition of water was called. For the benefit of students of that controversy it may be mentioned that the two "interpolations" of Cavendish's secretary, Sir Charles Blagden, and the addition made to the paper by Cavendish, after it was read and before it was printed, comprise the paragraph on page 20, beginning "All the foregoing experiments"; the paragraph on page 25, beginning "As Mr. Watt"; and the latter paragraphs of the paper from and including that on page 35, beginning "There are several memoirs of Mr. Lavoisier."

The second paper, which was published in 1785, contains the account of the discovery of the composition of nitric acid.

L. D.

EXPERIMENTS ON AIR.

FIRST PAPER.

Philosophical Transactions, Vol. 74 (for 1784), pp. 119-153.

Read January 15, 1784.

THE following experiments were made principally with a view to find out the cause of the diminution which common air is well known to suffer by all the various ways in which it is phlogisticated, and to discover what becomes of the air thus lost or condensed; and as they seem not only to determine this point, but also to throw great light on the constitution and manner of production of dephlogisticated air, I hope they may be not unworthy the acceptance of this society.

Many gentlemen have supposed that fixed air is either generated or separated from atmospheric air by phlogistication, and that the observed diminution is owing to this cause; my first experiments therefore were made in order to ascertain whether any fixed air is really produced thereby. Now, it must be observed, that as all animal and vegetable substances contain fixed air, and yield it by burning, distillation, or putrefaction, nothing can be concluded from experiments in which the air is phlogisticated by them. The only methods I know, which are not liable to objection, are by the calcination of metals, the burning of sulphur or phosphorus, the mixture of nitrous air, and the explosion of inflammable air. Perhaps it may be supposed, that I ought to add to these the electric spark; but I think it much most likely, that the phlogistication of the air, and production of fixed air, in this process, is owing to the burning of some inflammable matter in the apparatus. When the spark is taken from a solution of tournsol, the burning of the tournsol may produce this effect; when it is taken from lime-water, the burning of some foulness adhering to the tube, or perhaps of some inflammable matter contained in the lime, may have the same effect; and when quicksilver or metallic knobs are used, the calcination of them may contribute to the phlogistication of the air, though not to the production of fixed air.

There is no reason to think that any fixed air is produced by the first method of phlogistication. Dr. Priestley never found lime-water to become turbid by the calcination of metals over it * : Mr. Lavoisier also found only a very slight and scarce perceptible turbid appearance, without any precipitation, to take place when lime-water was shaken in a glass vessel full of the air in which lead had been calcined; and even this small diminution of transparency in the lime-water might very likely arise, not from fixed air, but only from its being fouled by particles of the calcined metal, which we are told adhered in some places to the glass. This want of turbidity has been attributed to the fixed air uniting to the metallic calx, in preference to the lime; but there is no reason for supposing that the calx contained any fixed air ; for I do not know that any one has extracted it from calces prepared in this manner ; and though most metallic calces prepared over the fire, or by long exposure to the atmosphere, where they are in contact with fixed air, contain that substance, it by no means follows that they must do so when prepared by methods in which they are not in contact with it.

Dr. Priestley also observed, that quicksilver, fouled by

---

* Experiments on Air, vol. I. p. 137.

the addition of lead or tin, deposits a powder by agitation and exposure to the air, which consists in great measure of the calx of the imperfect metal. He found too some powder of this kind to contain fixed air * ; but it is by no means clear that this air was produced by the phlogistication of the air in which the quicksilver was shaken ; as the powder was not prepared on purpose, but was procured from quicksilver fouled by having been used in various experiments, and may therefore have contained other impurities besides the metallic calces.

I never heard of any fixed air being produced by the burning of sulphur or phosphorus ; but it has been asserted, and commonly believed, that lime water is rendered cloudy by a mixture of common and nitrous air ; which, if true, would be a convincing proof that on mixing those two substances some fixed air is either generated or separated ; I therefore examined this carefully. Now it must be observed, that as common air usually contains a little fixed air, which is no essential part of it, but is easily separated by lime-water ; and as nitrous air may also contain fixed air, either if the metal from which it is procured be rusty, or if the water of the vessel in which it is caught contain calcareous earth, suspended by fixed air, as most waters do, it is proper first to free both airs from it by previously washing them with lime water †. Now I found, by repeated experi-

---

* Exper. in Nat. Phil. vol. I. p. 144.

† Though fixed air is absorbed in considerable quantity by water, as I shewed in Phil. Trans. vol. LVI. yet it is not easy to deprive common air of all the fixed air contained in it by means of water. On shaking a mixture of ten parts of common air, and one of fixed air, with more than an equal bulk of distilled water, not more than half of the fixed air was absorbed, and on transferring the air into fresh distilled water only half the remainder was absorbed, as appeared by the diminution which it still suffered on adding lime water.

ments, that if the lime water was clean, and the two airs were previously washed with that substance, not the least cloud was produced, either immediately on mixing them, or on suffering them to stand upwards of an hour, though it appeared by the thick clouds which were produced in the lime water, by breathing through it after the experiment was finished, that it was more than sufficient to saturate the acid formed by the decomposition of the nitrous air, and consequently that if any fixed air had been produced, it must have become visible. Once indeed I found a small cloud to be formed on the surface, after the mixture had stood a few minutes. In this experiment the lime water was not quite clean ; but whether the cloud was owing to this circumstance, or to the air's having not been properly washed, I cannot pretend to say.

Neither does any fixed air seem to be produced by the explosion of the inflammable air obtained from metals, with either common or dephlogisticated air. This I tried by putting a little lime-water into a glass globe fitted with a brass cock, so as to make it air tight, and an apparatus for firing air by electricity. This globe was exhausted by an air-pump, and the two airs, which had been previously washed with lime-water, let in, and suffered to remain some time, to shew whether they would affect the lime-water, and then fired by electricity. The event was, that not the least cloud was produced in the lime-water, when the inflammable air was mixed with common air, and only a very slight one, or rather diminution of transparency, when it was combined with dephlogisticated air. This, however, seemed not to be produced by fixed air ; as it appeared instantly after the explosion, and did not increase on standing, and was spread uniformly through the liquor ; whereas if it had been owing to fixed air, it would have taken up some short time before it appeared, and would have begun first at the surface, as was the case in the abovementioned experiment with nitrous air. What it was really owing to I cannot pretend to say ; but if it did proceed from fixed air it would shew that only an excessively minute quantity was produced *. On the whole, though it is not improbable that fixed air may be generated in some chymical processes, yet it seems certain that it is not the general effect of phlogisticating air, and that the diminution of common air is by no means owing to the generation or separation of fixed air from it.

As there seemed great reason to think, from Dr. Priestley's experiments, that the nitrous and vitriolic acids were convertible into dephlogisticated air, I tried whether the dephlogisticated part of common air might not, by phlogistication, be changed into nitrous or vitriolic acid. For this purpose I impregnated some milk of lime with the fumes of burning sulphur, by putting a little of it into a large glass receiver, and burning sulphur therein, taking care to keep the mouth of the receiver stopt till the fumes were all absorbed ; after which the air of the receiver was changed, and more sulphur burnt in it as before, and the process repeated till 122 grains of sulphur were consumed. The milk of lime was then filtered and evaporated, but it yielded no nitrous salt, nor any other substance except selenite ; so that no sensible quantity of the air was changed into nitrous acid. It must be observed, that as the vitriolic acid produced by the burning sulphur is changed by its union with the lime into selenite, which is very little soluble in water, a very small quantity of nitrous salt, or any other substance which is soluble in water, would have been perceived.

---

* Dr. Priestley also found no fixed air to be produced by the explosion of inflammable and common air. Vol. V. p. 124.

I also tried whether any nitrous acid was produced by phlogisticating common air with liver of sulphur; for this purpose I made a solution of flowers of sulphur by boiling it with lime, and put a little of it into a large receiver, and shook it frequently, changing now and then the air, till the yellow colour of the solution was quite gone; a sign that all the sulphur was, by the loss of its phlogiston, turned into vitriolic acid, and united to the lime, or precipitated; the liquor was then filtered and evaporated, but it yielded not the least nitrous salt.

The experiment was repeated in nearly the same manner with dephlogisticated air procured from red precipitate; but not the least nitrous acid was obtained.

It is well known that common selenite is very little soluble in water; whereas that procured in the two last experiments was very soluble, and even crystallized readily, and was intensely bitter; this however appeared to be owing merely to the acid with which it was formed being very much phlogisticated; for on evaporating it to dryness, and exposing it to the air for a few days, it became much less soluble, so that on adding water to it not much dissolved; and by repeating this process once or twice, it seemed to become not more soluble than selenite made in the common manner.

This solubility of the selenite caused some trouble in trying the experiment; for while it continued much soluble it would have been impossible to have distinguished a small mixture of nitrous salt; but by the abovementioned process I was able to distinguish as small a proportion as if the selenite had been originally no more soluble than usual.

The nature of the neutral salts made with the phlogisticated vitriolic and nitrous acids has not been much examined by the chymists, though it seems well worth their attention; and it is likely that many besides the foregoing may differ remarkably from those made with the same acids in their common state. Nitre formed with the phlogisticated nitrous acid has been found to differ considerably from common nitre, as well as Sal Polychrest from vitriolated tartar.

In order to try whether any vitriolic acid was produced by the phlogistication of air, I impregnated fifty ounces of distilled water with the fumes produced on mixing fifty-two ounce measures of nitrous air with a quantity of common air sufficient to decompound it. This was done by filling a bottle with some of this water, and inverting it into a bason of the same, and then, by a syphon, letting in as much nitrous air as filled it half-full; after which common air was added slowly by the same syphon, till all the nitrous air was decompounded. When this was done, the distilled water was further impregnated in the same manner till the whole of the above-mentioned quantity of nitrous air was employed. This impregnated water, which was very sensibly acid to the taste, was distilled in a glass retort. The first runnings were very acid, and smelt pungent, being nitrous acid much phlogisticated; what came next had no sensible taste or smell; but the last runnings were very acid, and consisted of nitrous acid not phlogisticated. Scarce any sediment was left behind. These different parcels of distilled liquor were then exactly saturated with salt of tartar, and evaporated; they yielded 87½ grains of nitre, which, as far as I could perceive, was unmixed with vitriolated tartar or any other substance, and consequently no sensible quantity of the common air with which the nitrous air was mixed was turned into vitriolic acid.

It appears, from this experiment, that nitrous air contains as much acid as 2¾ times its weight of saltpetre; for fifty-two ounce measures of nitrous air weigh 32 grains, and, as was before said, yield as much acid as is contained in 87½ grains of saltpetre ; so that the acid in nitrous air is in a remarkably concentrated state, and I believe more than 1½ times as much so as the strongest spirit of nitre ever prepared.

Having now mentioned the unsuccessful attempts I made to find out what becomes of the air lost by phlogistication, I proceed to some experiments, which serve really to explain the matter.

In Dr. Priestley's last volume of experiments is related an experiment of Mr. Warltire's, in which it is said that, on firing a mixture of common and inflammable air by electricity in a close copper vessel holding about three pints, a loss of weight was always perceived, on an average about two grains, though the vessel was stopped in such a manner that no air could escape by the explosion. It is also related, that on repeating the experiment in glass vessels, the inside of the glass, though clean and dry before, immediately became dewy ; which confirmed an opinion he had long entertained, that common air deposits its moisture by phlogistication. As the latter experiment seemed likely to throw great light on the subject I had in view, I thought it well worth examining more closely. The first experiment also, if there was no mistake in it, would be very extraordinary and curious ; but it did not succeed with me ; for though the vessel I used held more than Mr. Warltire's, namely, 24,000 grains of water, and though the experiment was repeated several times with different proportions of common and inflammable air, I could never perceive a loss of weight of more than one-fifth of a grain, and commonly none at all. It must be observed, however, that though there were some of the experiments in which it seemed to diminish a little in weight, there were none in which it increased *.

---

* Dr. Priestley, I am informed, has since found the experiment not to succeed.

In all the experiments, the inside of the glass globe became dewy, as observed by Mr. Warltire; but not the least sooty matter could be perceived. Care was taken in all of them to find how much the air was diminished by the explosion, and to observe its test. The result is as follows: the bulk of the inflammable air being expressed in decimals of the common air,

Common air.	Inflammable air.	Diminution.	Air remaining after the explosion.	Test of this air in first method.	Standard.
	1,241	,686	1,555	,055	,0
1	1,055	,642	1,413	,063	,0
	,706	,647	1,059	,066	,0
	,423	,612	,811	,097	,03
	,331	,476	,855	,339	,27
	,206	,294	,912	,648	,58

In these experiments the inflammable air was procured from zinc, as it was in all my experiments, except where otherwise expressed: but I made two more experiments, to try whether there was any difference between the air from zinc and that from iron, the quantity of inflammable air being the same in both, namely, 0,331 of the common; but I could not find any difference to be depended on between the two kinds of air, either in the diminution which they suffered by the explosion, or the test of the burnt air.

From the fourth experiment it appears, that 423 measures of inflammable air are nearly sufficient to completely phlogisticate 1000 of common air; and that the bulk of the air remaining after the explosion is then very little more than four-fifths of the common air employed; so that as common air cannot be reduced to a much less bulk than that by any method of phlogistication, we may safely conclude, that when they are mixed in this proportion, and exploded, almost all the inflammable air, and about one-fifth part of the common air, lose their elasticity, and are condensed into the dew which lines the glass.

The better to examine the nature of this dew, 500000 grain measures of inflammable air were burnt with about $2\frac{1}{2}$ times that quantity of common air, and the burnt air made to pass through a glass cylinder eight feet long and three-quarters of an inch in diameter, in order to deposit the dew. The two airs were conveyed slowly into this cylinder by separate copper pipes, passing through a brass plate which stopped up the end of the cylinder; and as neither inflammable nor common air can burn by themselves, there was no danger of the flame spreading into the magazines from which they were conveyed. Each of these magazines consisted of a large tin vessel, inverted into another vessel just big enough to receive it. The inner vessel communicated with the copper pipe, and the air was forced out of it by pouring water into the outer vessel; and in order that the quantity of common air expelled should be $2\frac{1}{2}$ times that of the inflammable, the water was let into the outer vessels by two holes in the bottom of the same tin pan, the hole which conveyed the water into that vessel in which the common air was confined being $2\frac{1}{2}$ times as big as the other.

In trying the experiment, the magazines being first filled with their respective airs, the glass cylinder was taken off, and water let, by the two holes, into the outer vessels, till the airs began to issue from the ends of the copper pipes; they were then set on fire by a candle, and the cylinder put on again in its place. By this means upwards of 135 grains of water were condensed in the cylinder, which had no taste nor smell, and which left no sensible sediment when evaporated to dryness; neither did it yield any pungent smell during the evaporation; in short, it seemed pure water.

In my first experiment, the cylinder near that part

where the air was fired was a little tinged with sooty matter, but very slightly so; and that little seemed to proceed from the putty with which the apparatus was luted, and which was heated by the flame; for in another experiment, in which it was contrived so that the luting should not be much heated, scarce any sooty tinge could be perceived.

By the experiments with the globe it appeared, that when inflammable and common air are exploded in a proper proportion, almost all the inflammable air, and near one-fifth of the common air, lose their elasticity, and are condensed into dew. And by this experiment it appears, that this dew is plain water, and consequently that almost all the inflammable air, and about one-fifth of the common air, are turned into pure water.

In order to examine the nature of the matter condensed on firing a mixture of dephlogisticated and inflammable air, I took a glass globe, holding 8800 grain measures, furnished with a brass cock and an apparatus for firing air by electricity. This globe was well exhausted by an air-pump, and then filled with a mixture of inflammable and dephlogisticated air, by shutting the cock, fastening a bent glass tube to its mouth, and letting up the end of it into a glass jar inverted into water, and containing a mixture of 19500 grain measures of dephlogisticated air, and 37000 of inflammable; so that, upon opening the cock, some of this mixed air rushed through the bent tube, and filled the globe*. The cock was then shut, and the included air fired by electricity, by which means almost all of it lost its elasti-

---

* In order to prevent any water from getting into this tube, while dipped under water to let it up into the glass jar, a bit of wax was stuck upon the end of it, which was rubbed off when raised above the surface of the water.

city. The cock was then again opened, so as to let in more of the same air, to supply the place of that destroyed by the explosion, which was again fired, and the operation continued till almost the whole of the mixture was let into the globe and exploded. By this means, though the globe held not more than the sixth part of the mixture, almost the whole of it was exploded therein, without any fresh exhaustion of the globe.

As I was desirous to try the quantity and test of this burnt air, without letting any water into the globe, which would have prevented my examining the nature of the condensed matter, I took a larger globe, furnished also with a stop cock, exhausted it by an air-pump, and screwed it on upon the cock of the former globe ; upon which, by opening both cocks, the air rushed out of the smaller globe into the larger, till it became of equal density in both ; then, by shutting the cock of the larger globe, unscrewing it again from the former, and opening it under water, I was enabled to find the quantity of the burnt air in it; and consequently, as the proportion which the contents of the two globes bore to each other was known, could tell the quantity of burnt air in the small globe before the communication was made between them. By this means the whole quantity of the burnt air was found to be 2950 grain measures ; its standard was 1,85.

The liquor condensed in the globe, in weight about 30 grains, was sensibly acid to the taste, and by saturation with fixed alkali, and evaporation, yielded near two grains of nitre ; so that it consisted of water united to a small quantity of nitrous acid. No sooty matter was deposited in the globe. The dephlogisticated air used in this experiment was procured from red precipitate, that is, from a solution of quicksilver in spirit of nitre dis-tilled till it acquires a red colour.

As it was suspected, that the acid contained in the condensed liquor was no essential part of the dephlogisticated air, but was owing to some acid vapour which came over in making it and had not been absorbed by the water, the experiment was repeated in the same manner, with some more of the same air, which had been previously washed with water, by keeping it a day or two in a bottle with some water, and shaking it frequently; whereas that used in the preceding experiment had never passed through water, except in preparing it. The condensed liquor was still acid.

The experiment was also repeated with dephlogisticated air, procured from red lead by means of oil of vitriol; the liquor condensed was acid, but by an accident I was prevented from determining the nature of the acid.

I also procured some dephlogisticated air from the leaves of plants, in the manner of Doctors Ingenhousz and Priestley, and exploded it with inflammable air as before; the condensed liquor still continued acid, and of the nitrous kind.

In all these experiments the proportion of inflammable air was such, that the burnt air was not much phlogisticated; and it was observed, that the less phlogisticated it was, the more acid was the condensed liquor. I therefore made another experiment, with some more of the same air from plants, in which the proportion of inflammable air was greater, so that the burnt air was almost completely phlogisticated, its standard being $\frac{1}{10}$. The condensed liquor was then not at all acid, but seemed pure water: so that it appears, that with this kind of dephlogisticated air, the condensed liquor is not at all acid, when the two airs are mixed in such a proportion that the burnt air is almost completely phlogisticated, but is considerably so when it is not much phlogisticated.

In order to see whether the same thing would obtain with air procured from red precipitate, I made two more experiments with that kind of air, the air in both being taken from the same bottle, and the experiment tried in the same manner, except that the proportions of inflammable air were different. In the first, in which the burnt air was almost completely phlogisticated, the condensed liquor was not at all acid. In the second, in which its standard was 1,86, that is, not much phlogisticated, it was considerably acid ; so that with this air, as well as with that from plants, the condensed liquor contains, or is entirely free from, acid, according as the burnt air is less or more phlogisticated ; and there can be little doubt but that the same rule obtains with any other kind of dephlogisticated air.

In order to see whether the acid, formed by the explosion of dephlogisticated air obtained by means of the vitriolic acid, would also be of the nitrous kind, I procured some air from turbith mineral, and exploded it with inflammable air, the proportion being such that the burnt air was not much phlogisticated. The condensed liquor manifested an acidity, which appeared, by saturation with a solution of salt of tartar, to be of the nitrous kind ; and it was found, by the addition of some terra ponderosa salita, to contain little or no vitriolic acid.

When inflammable air was exploded with common air, in such a proportion that the standard of the burnt air was about $\frac{4}{10}$, the condensed liquor was not in the least acid. There is no difference, however, in this respect between common air, and dephlogisticated air mixed with phlogisticated in such a proportion as to reduce it to the standard of common air ; for some dephlogisticated air from red precipitate, being reduced to this standard by the addition of perfectly phlogisticated air, and then exploded with the same proportion of inflammable air as the common air was in the foregoing experiment, the condensed liquor was not in the least acid.

From the foregoing experiments it appears, that when a mixture of inflammable and dephlogisticated air is exploded in such proportion that the burnt air is not much phlogisticated, the condensed liquor contains a little acid, which is always of the nitrous kind, whatever substance the dephlogisticated air is procured from ; but if the proportion be such that the burnt air is almost entirely phlogisticated, the condensed liquor is not at all acid, but seems pure water, without any addition whatever ; and as, when they are mixed in that proportion, very little air remains after the explosion, almost the whole being condensed, it follows, that almost the whole of the inflammable and dephlogisticated air is converted into pure water. It is not easy, indeed, to determine from these experiments what proportion the burnt air, remaining after the explosions, bore to the dephlogisticated air employed, as neither the small nor the large globe could be perfectly exhausted of air, and there was no saying with exactness what quantity was left in them ; but in most of them, after allowing for this uncertainty, the true quantity of burnt air seemed not more than $\frac{1}{17}$th of the dephlogisticated air employed, or $\frac{1}{18}$th of the mixture. It seems, however, unnecessary to determine this point exactly, as the quantity is so small, that there can be little doubt but that it proceeds only from the impurities mixed with the dephlogisticated and inflammable air, and consequently that, if those airs could be obtained perfectly pure, the whole would be condensed.

With respect to common air, and dephlogisticated air reduced by the addition of phlogisticated air to the standard of common air, the case is different ; as the liquor condensed in exploding them with inflammable air, I believe I may say in any proportion, is not at all acid ; perhaps, because if they are mixed in such a proportion as that the burnt air is not much phlogisticated, the explosion is too weak, and not accompanied with sufficient heat.

All the foregoing experiments, on the explosion of inflammable air with common and dephlogisticated airs, except those which relate to the cause of the acid found in the water, were made in the summer of the year 1781, and were mentioned by me to Dr. Priestley, who in consequence of it made some experiments of the same kind, as he relates in a paper printed in the preceding volume of the Transactions. During the last summer also, a friend of mine gave some account of them to M. Lavoisier, as well as of the conclusion drawn from them, that dephlogisticated air is only water deprived of phlogiston ; but at that time so far was M. Lavoisier from thinking any such opinion warranted, that, till he was prevailed upon to repeat the experiment himself, he found some difficulty in believing that nearly the whole of the two airs could be converted into water. It is remarkable, that neither of these gentlemen found any acid in the water produced by the combustion ; which might proceed from the latter having burnt the two airs in a different manner from what I did : and from the former having used a different kind of inflammable air, namely, that from charcoal, and perhaps having used a greater proportion of it.

Before I enter into the cause of these phænomena, it will be proper to take notice, that phlogisticated air appears to be nothing else than the nitrous acid united to phlogiston ; for when nitre is deflagrated with charcoal, the acid is almost entirely converted into this kind of air. That the acid is entirely converted into air, appears from the common process for making what is called clyssus of nitre ; for if the nitre and charcoal are dry, scarce anything is found in the vessels prepared for condensing the fumes; but if they are moist a little liquor is collected, which is nothing but the water contained in the materials, impregnated with a little volatile alkali, proceeding in all probability from the imperfectly burnt charcoal, and a little fixed alkali, consisting of some of the alkalized nitre carried over by the heat and watery vapours. As far as I can perceive too, at present, the air into which much the greatest part of the acid is converted, differs in no respect from common air phlogisticated. A small part of the acid, however, is turned into nitrous air, and the whole is mixed with a good deal of fixed, and perhaps a little inflammable air, both proceeding from the charcoal.

It is well known, that the nitrous acid is also converted by phlogistication into nitrous air, in which respect there seems a considerable analogy between that and the vitriolic acid; for the vitriolic acid, when united to a smaller proportion of phlogiston, forms the volatile sulphureous acid and vitriolic acid air, both of which, by exposure to the atmosphere, lose their phlogiston, though not very fast, and are turned back into vitriolic acid; but, when united to a greater proportion of phlogiston, it forms sulphur, which shews no signs of acidity, unless a small degree of affinity to alkalies can be called so, and in which the phlogiston is more strongly adherent, so that it does not fly off when exposed to the air, unless assisted by a heat sufficient to set it on fire. In like manner the nitrous acid, united to a certain quantity of phlogiston, forms nitrous fumes and nitrous air, which readily quit their phlogiston to common air; but when united to a different, in all probability a larger quantity, it forms phlogisticated air, which shews no signs of acidity, and is still less disposed to part with its phlogiston than sulphur.

This being premised, there seem two ways by which the phænomena of the acid found in the condensed liquor may be explained; first, by supposing that dephlogisticated air contains a little nitrous acid which enters into it as one of its component parts, and that this acid, when the inflammable air is in a sufficient proportion, unites to the phlogiston, and is turned into phlogisticated air, but does not when the inflammable air is in too small a proportion; and, secondly, by supposing that there is no nitrous acid mixed with, or entering into the composition of, dephlogisticated air, but that, when this air is in a sufficient proportion, part of the phlogisticated air with which it is debased is, by the strong affinity of phlogiston to dephlogisticated air, deprived of its phlogiston and turned into nitrous acid; whereas, when the dephlogisticated air is not more than sufficient to consume the inflammable air, none then remains to deprive the phlogisticated air of its phlogiston, and turn it into acid.

If the latter explanation be true, I think, we must allow that dephlogisticated air is in reality nothing but dephlogisticated water, or water deprived of its phlogiston; or, in other words, that water consists of dephlogisticated air united to phlogiston; and that inflammable air is either pure phlogiston, as Dr. Priestley and Mr. Kirwan suppose, or else water united to phlogiston * ;

---

* Either of these suppositions will agree equally well with the following experiments; but the latter seems to me much the most likely. What principally makes me think so is, that common or dephlogisticated air do not absorb phlogiston from inflammable air, unless assisted by a red heat, whereas they absorb the phlogiston of nitrous air, liver of sulphur, and many other substances, without that assistance; and it seems inexplicable, that they should refuse to unite to pure phlogiston, when they are able to extract it from substances to which it has an affinity : that is, that they should overcome the affinity of phlogiston to other substances, and extract

since, according to this supposition, these two substances united together form pure water. On the other hand, if the first explanation be true, we must suppose that dephlogisticated air consists of water united to a little nitrous acid and deprived of its phlogiston ; but still the nitrous acid in it must make only a very small part of the whole, as it is found, that the phlogisticated air, which it is converted into, is very small in comparison of the dephlogisticated air.

I think the second of these explanations seems much the most likely ; as it was found, that the acid in the condensed liquor was of the nitrous kind, not only when the dephlogisticated air was prepared from red precipitate, but also when it was procured from plants or from turbith mineral : and it seems not likely, that air procured from plants, and still less likely that air procured from a solution of mercury in oil of vitriol, should contain any nitrous acid.

Another strong argument in favour of this opinion is, that dephlogisticated air yields no nitrous acid when phlogisticated by liver of sulphur ; for if this air contains nitrous acid, and yields it when phlogisticated by explosion with inflammable air, it is very extraordinary that it should not do so when phlogisticated by other means.

it from them, when they will not even unite to it when presented to them. On the other hand, I know no experiment which shews inflammable air to be pure phlogiston rather than an union of it with water, unless it be Dr. Priestley's experiment of expelling inflammable air from iron by heat alone. I am not sufficiently acquainted with the circumstances of that experiment to argue with certainty about it ; but I think it much more likely, that the inflammable air was formed by the union of the phlogiston of the iron filings with the water dispersed among them, or contained in the retort or other vessel in which it was heated ; and in all probability this was the cause of the separation of the phlogiston, as iron seems not disposed to part with its phlogiston by heat alone, without being assisted by the air or some other substance.

But what forms a stronger and, I think, almost decisive argument in favour of this explanation is, that when the dephlogisticated air is very pure, the condensed liquor is made much more strongly acid by mixing the air to be exploded with a little phlogisticated air, as appears by the following experiments.

A mixture of 18500 grain measures of inflammable air with 9750 of dephlogisticated air procured from red precipitate were exploded in the usual manner; after which, a mixture of the same quantities of the same dephlogisticated and inflammable air, with the addition of 2500 of air phlogisticated by iron filings and sulphur, was treated in the same manner. The condensed liquor, in both experiments, was acid, but that in the latter evidently more so, as appeared also by saturating each of them separately with marble powder, and precipitating the earth by fixed alkali, the precipitate of the second experiment weighing one-fifth of a grain, and that of the first being several times less. The standard of the burnt air in the first experiment was 1,86, and in the second only 0,9.

It must be observed, that all circumstances were the same in these two experiments, except that in the latter the air to be exploded was mixed with some phlogisticated air, and that in consequence the burnt air was more phlogisticated than in the former; and from what has been before said, it appears, that this latter circumstance ought rather to have made the condensed liquor less acid; and yet it was found to be much more so, which shews strongly that it was the phlogisticated air which furnished the acid.

As a further confirmation of this point, these two comparative experiments were repeated with a little variation, namely, in the first experiment there was first let into the globe 1500 of dephlogisticated air, and then the mixture, consisting of 12200 of dephlogisticated air and 25900 of inflammable, was let in at different times as usual. In the second experiment, besides the 1500 of dephlogisticated air first let in, there was also admitted 2500 of phlogisticated air, after which the mixture, consisting of the same quantities of dephlogisticated and inflammable air as before, was let in as usual. The condensed liquor of the second experiment was about three times as acid as that of the first, as it required 119 grains of a diluted solution of salt of tartar to saturate it, and the other only 37. The standard of the burnt air was 0,78 in the second experiment, and 1,96 in the first.

The intention of previously letting in some dephlogisticated air in the two last experiments was, that the condensed liquor was expected to become more acid thereby, as proved actually to be the case.

In the first of these two experiments, in order that the air to be exploded should be as free as possible from common air, the globe was first filled with a mixture of dephlogisticated and inflammable air, it was then exhausted, and the air to be exploded let in; by which means, though the globe was not perfectly exhausted, very little common air could be left in it. In the first set of experiments this circumstance was not attended to, and the purity of the dephlogisticated air was forgot to be examined in both sets.

From what has been said there seems the utmost reason to think, that dephlogisticated air is only water deprived of its phlogiston, and that inflammable air, as was before said, is either phlogisticated water, or else pure phlogiston; but in all probability the former.

As Mr. Watt, in a paper lately read before this Society, supposes water to consist of dephlogisticated air and phlogiston deprived of part of their latent heat, whereas I take no notice of the latter circumstance, it may be proper to mention in a few words the reason of this apparent difference between us. If there be any such thing as elementary heat, it must be allowed that what Mr. Watt says is true; but by the same rule we ought to say, that the diluted mineral acids consist of the concentrated acids united to water and deprived of part of their latent heat; that solutions of sal ammoniac, and most other neutral salts, consist of the salt united to water and elementary heat; and a similar language ought to be used in speaking of almost all chemical combinations, as there are very few which are not attended with some increase or diminution of heat. Now I have chosen to avoid this form of speaking, both because I think it more likely that there is no such thing as elementary heat, and because saying so in this instance, without using similar expressions in speaking of other chemical unions, would be improper, and would lead to false ideas; and it may even admit of doubt, whether the doing it in general would not cause more trouble and perplexity than it is worth.

There is the utmost reason to think, that dephlogisticated and phlogisticated air, as M. Lavoisier and Scheele suppose, are quite distinct substances, and not differing only in their degree of phlogistication; and that common air is a mixture of the two; for if the dephlogisticated air is pretty pure, almost the whole of it loses its elasticity by phlogistication, and, as appears by the foregoing experiments, is turned into water, instead of being converted into phlogisticated air. In most of the foregoing experiments, at least $\frac{4}{5}$ths of the whole was turned into water; and by treating some dephlogisticated air with liver of sulphur, I have reduced it to less than $\frac{1}{80}$th of its original bulk, and other persons, I believe, have reduced it to a still less bulk: so that there seems the utmost reason to suppose, that the small residuum which remains after its phlogistication proceeds only from the impurities mixed with it.

It was just said, that some dephlogisticated air was reduced by liver of sulphur to $\frac{1}{30}$th of its original bulk; the standard of this air was 4,8, and consequently the standard of perfectly pure dephlogisticated air should be very nearly 5, which is a confirmation of the foregoing opinion; for if the standard of pure dephlogisticated air is 5, common air must, according to this opinion, contain one-fifth of it, and therefore ought to lose one-fifth of its bulk by phlogistication, which is what it is actually found to lose.

From what has been said, it follows, that instead of saying air is phlogisticated or dephlogisticated by any means, it would be more strictly just to say, it is deprived of, or receives, an addition of dephlogisticated air; but as the other expression is convenient, and can scarcely be considered as improper, I shall still frequently make use of it in the remainder of this paper.

There seemed great reason to think, from Dr. Priestley's experiments, that both the nitrous and vitriolic acids were convertible into dephlogisticated air, as that air is procured in the greatest quantity from substances containing those acids, especially the former. The foregoing experiments, however, seem to shew that no part of the acid is converted into dephlogisticated air, and that their use in preparing it is owing only to the great power which they possess of depriving bodies of their phlogiston. A strong confirmation of this is, that red precipitate, which is one of the substances yielding dephlogisticated air in the greatest quantity, and which is prepared by means of the nitrous acid, contains in reality no acid. This I found by grinding 400 grains of it with spirits of sal ammoniac, and keeping them together for some days in a bottle, taking care to shake them frequently. The red colour of the precipitate was rendered pale, but not entirely destroyed; being then washed with water and filtered, the clear liquor yielded on evaporation not the least ammoniacal salt.

It is natural to think, that if any nitrous acid had been contained in the red precipitate, it would have united to the volatile alkali and have formed ammoniacal nitre, and would have been perceived on evaporation; but in order to determine more certainly whether this would be the case, I dried some of the same solution of quicksilver from which the red precipitate was prepared with a less heat, so that it acquired only an orange colour, and treated the same quantity of it with volatile alkali in the same manner as before. It immediately caused an effervescence, changed the colour to grey, and yielded 52 grains of ammoniacal nitre. There is the utmost reason to think, therefore, that red precipitate contains no nitrous acid; and consequently that, in procuring dephlogisticated air from it, no acid is converted into air; and it is reasonable to conclude, therefore, that no such change is produced in procuring it from any other substance.

It remains to consider in what manner these acids act in producing dephlogisticated air. The way in which the nitrous acid acts, in the production of it from red precipitate, seems to be as follows. On distilling the mixture of quicksilver and spirit of nitre, the acid comes over, loaded with phlogiston, in the form of nitrous vapour, and continues to do so till the remaining matter acquires its full red colour, by which time all the nitrous acid is driven over, but some of the watery part still remains behind, and adheres strongly to the quicksilver; so that the red precipitate may be considered, either as quicksilver deprived of part of its phlogiston, and united to a certain portion of water, or as quicksilver united to dephlogisticated air * ; after which, on further increasing the heat, the water in it rises deprived of its phlogiston, that is, in the form of dephlogisticated air, and at the same time the quicksilver distils over in its metallic form. It is justly remarked by Dr. Priestley, that the solution of quicksilver does not begin to yield dephlogisticated air till it acquires its red colour.

Mercurius calcinatus appears to be only quicksilver which has absorbed dephlogisticated air from the atmosphere during its preparation ; accordingly, by giving it a sufficient heat, the dephlogisticated air is driven off, and the quicksilver acquires its original form. It seems therefore that mercurius calcinatus and red precipitate, though prepared in a different manner, are very nearly the same thing.

From what has been said it follows, that red precipitate and mercurius calcinatus contain as much phlogiston as the quicksilver they are prepared from ; but yet, as uniting dephlogisticated air to a metal comes to the same thing as depriving it of part of its phlogiston and adding water to it, the quicksilver may still be considered as deprived of its phlogiston ; but the imperfect metals seem not only to absorb dephlogisticated air during their calcination, but also to be really deprived of part of their

---

* Unless we were much better acquainted than we are with the manner in which different substances are united together in compound bodies, it would be ridiculous to say, that it is the quicksilver in the red precipitate which is deprived of its phlogiston, and not the water, or that it is the water and not the quicksilver ; all that we can say is, that red precipitate consists of quicksilver and water, one or both of which are deprived of part of their phlogiston. In like manner, during the preparation of the red precipitate, it is certain that the acid absorbs phlogiston, either from the quicksilver or the water ; but we are by no means authorised to say from which.

phlogiston, as they do not acquire their metallic form by driving off the dephlogisticated air.

In procuring dephlogisticated air from nitre, the acid acts in a different manner, as, upon heating the nitre red-hot, the dephlogisticated air rises mixed with a little nitrous acid, and at the same time the acid remaining in the nitre becomes very much phlogisticated; which shews that the acid absorbs phlogiston from the water in the nitre, and becomes phlogisticated, while the water is thereby turned into dephlogisticated air. On distilling 3155 grains of nitre in an unglazed earthen retort, it yielded 256000 grain measures of dephlogisticated air *, the standard of different parts of which varied from 3 to 3,65, but at a medium was 3,35. The matter remaining in the retort dissolved readily in water, and tasted alcaline and caustic. On adding diluted spirit of nitre to the solution, strong red fumes were produced; a sign that the acid in it was very much phlogisticated, as no fumes whatever would have been produced on adding the same acid to a solution of common nitre : that part of the solution also which was supersaturated with acid became blue ; a colour which the diluted nitrous acid is known to assume when much phlogisticated. The solution, when saturated with this acid, lost its alcaline and caustic taste, but yet tasted very different from true nitre, seeming as if it had been mixed with sea-salt, and also required much less water to dissolve it ; but on exposing it for some days to the air, and adding fresh acid as fast as by the flying off of the fumes the alcali

---

* This is, about eighty-one grain measures from one grain of nitre ; and the weight of the dephlogisticated air, supposing it 800 times lighter than water, is one tenth of that of the nitre. In all probability it would have yielded a much greater quantity of air, if a greater heat had been applied.

predominated, it became true nitre, unmixed, as far as I could perceive, with any other salt *.

It has been remarked, that the dephlogisticated air procured from nitre is less pure, than that from red precipitate and many other substances, which may perhaps proceed from unglazed earthen retorts having been commonly used for this purpose, and which, conformably to Dr. Priestley's discovery, may possibly absorb some common air from without, and emit it along with the dephlogisticated air ; but if it should be found that the dephlogisticated air procured from nitre in glass or glazed earthen vessels is also impure, it would seem to shew that part of the acid in the nitre is turned into phlogisticated air, by absorbing phlogiston from the watery part.

From what has been said it appears, that there is a considerable difference in the manner in which the acid acts in the production of dephlogisticated air from red precipitate and from nitre ; in the former case the acid comes over first, leaving the remaining substance deprived of part of its phlogiston ; in the latter the dephlogisticated air comes first, leaving the acid loaded with the phlogiston of the water from which it was formed.

On distilling a mixture of quicksilver and oil of vitriol to dryness, part of the acid comes over, loaded with phlogiston, in the form of volatile sulphureous acid and vitriolic acid air ; so that the remaining white mass may be considered as consisting of quicksilver deprived of its phlogiston, and united to a certain proportion of acid and water, or of plain quicksilver united to a certain proportion of acid and dephlogisticated air.

---

* This phlogistication of the acid in nitre by heat has been observed by Mr. Scheele ; see his experiments on air and fire, p. 45. English translation.

Accordingly on urging this white mass with a more violent heat, the dephlogisticated air comes over, and at the same time part of the quicksilver rises in its metallic form, and also part of the white mass, united in all probability to a greater proportion of acid than before, sublimes ; so that the rationale of the production of dephlogisticated air from turbith mineral, and from red precipitate, are nearly similar.

True turbith mineral consists of the abovementioned white mass, well washed with water, by which means it acquires a yellow colour, and contains much less acid than the unwashed mass. Accordingly it seems likely, that on exposing this to heat, less of it should sublime without being decompounded, and consequently that more dephlogisticated air should be procured from it than from the unwashed mass.

This is an instance, that the superabundant vitriolic acid may, in some cases, be better extracted from the base it is united to by water than by heat. Vitriolated tartar is another instance ; for, if vitriolated tartar be mixed with oil of vitriol and exposed even to a pretty strong red heat, the mass will be very acid ; but, if this mass is dissolved in water, and evaporated, the crystals will be not sensibly so.

In all probability, the vitriolic acid acts in the same manner in the production of dephlogisticated air from alum, as the nitrous does in its production from nitre ; that is, the watery part comes over first in the form of dephlogisticated air, leaving the acid charged with its phlogiston. Whether this is also the case with regard to green and blue vitriol, or whether in them the acid does not rather act in the same manner as in turbith mineral, I cannot pretend to say, but I think the latter more likely.

There is another way by which dephlogisticated air

has been found to be produced in great quantities, namely, the growth of vegetables exposed to the sun or day-light ; the rationale of which, in all probability, is, that plants, when assisted by the light, deprive part of the water sucked up by their roots of its phlogiston, and turn it into dephlogisticated air, while the phlogiston unites to, and forms part of, the substance of the plant.

There are many circumstances which shew, that light has a remarkable power in enabling one body to absorb phlogiston from another. Mr. Senebier has observed, that the green tincture procured from the leaves of vegetables by spirit of wine, quickly loses its colour when exposed to the sun in a bottle not more than one third part full, but does not do so in the dark, or if the bottle is quite full of the tincture, or if the air in it is phlogisticated ; whence it is natural to conclude, that the light enables the dephlogisticated part of the air to absorb phlogiston from the tincture ; and this appears to be really the case, as I find that the air in the bottle is considerably phlogisticated thereby. Dephlogisticated spirit of nitre also acquires a yellow colour, and becomes phlogisticated, by exposure to the sun's rays * ; and I find on trial that the air in the bottle in which it is contained becomes dephlogisticated, or, in other words, receives an increase of dephlogisticated air, which shews that the change in the acid is not owing to the sun's rays

---

* If spirit of nitre is distilled with a very gentle heat, the part which comes over is high coloured and fuming, and that which remains behind is quite colourless, and fumes much less than other nitrous acid of the same strength, and the fumes are colourless. This is called dephlogisticated spirit of nitre, as it appears to be really deprived of phlogiston by the process. The manner of preparing it, as well as its property of regaining its yellow colour by exposure to the light, is mentioned by Mr. Scheele in the Stockholm Memoirs, 1774.

communicating phlogiston to it, but to their enabling it to absorb phlogiston from the water contained in it, and thereby to produce dephlogisticated air. Mr. Scheele also found, that the dark colour acquired by luna cornea on exposure to the light, is owing to part of the silver being revived ; and that gold, dissolved in aqua regia and deprived by distillation of the nitrous and super-fluous marine acid, is revived by the same means ; and there is the utmost reason to think, that, in both cases, the revival of the metal is owing to its absorbing phlo-giston from the water.

Vegetables seem to consist almost intirely of fixed and phlogisticated air, united to a large proportion of phlo-giston and some water, since by burning in the open air, in which their phlogiston unites to the dephlogisticated part of the atmosphere and forms water, they seem to be reduced almost intirely to water and those two kinds of air. Now plants growing in water without earth, can receive nourishment only from the water and air, and must therefore in all probability absorb their phlogiston from the water. It is known also that plants growing in the dark do not thrive well, and grow in a very different manner from what they do when exposed to the light.

From what has been said, it seems likely that the use of light, in promoting the growth of plants and the pro-duction of dephlogisticated air from them, is, that it en-ables them to absorb phlogiston from the water. To this it may perhaps be objected, that though plants do not thrive well in the dark, yet they do grow, and should there-fore, according to this hypothesis, absorb water from the atmosphere, and yield dephlogisticated air, which they have not been found to do. But we have no proof that they grew at all in any of those cases in which they were found not to yield dephlogisticated air ; for though they will grow in the dark, yet their vegetative powers may perhaps at first be intirely checked by it, especially considering the unnatural situation in which they must be placed in such experiments. Perhaps too plants growing in the dark may be able to absorb phlogiston from water not much impregnated with dephlogisticated air, but not from water strongly impregnated with it ; and consequently, when kept under water in the dark, may perhaps at first yield some dephlogisticated air, which, instead of rising to the surface, may be absorbed by the water, and, before the water is so much impregnated as to suffer any to escape, the plant may cease to vegetate, unless the water is changed. Unless therefore it could be shown that plants growing in the dark, in water alone, will increase in size, without yielding dephlogisticated air, and without the water becoming more impregnated with it than before, no objection can be drawn from thence.

Mr. Senebier finds, that plants yield much more dephlogisticated air in distilled water impregnated with fixed air, than in plain distilled water, which is perfectly conformable to the abovementioned hypothesis ; for as fixed air is a principal constituent part of vegetable substances, it is reasonable to suppose that the work of vegetation will go on better in water containing this substance, than in other water.

There are several memoirs of Mr. Lavoisier published by the Academy of Sciences, in which he intirely discards phlogiston, and explains those phænomena which have been usually attributed to the loss or attraction of that substance, by the absorption or expulsion of dephlogisticated air ; and as not only the foregoing experiments, but most other phænomena of nature, seem explicable as well, or nearly as well, upon this as upon the commonly believed principle of phlogiston, it may be proper briefly to mention in what manner I would explain them on this principle, and why I have adhered to the other. In doing this, I shall not conform strictly to his theory, but shall make such additions and alterations as seem to suit it best to the phænomena; the more so, as the foregoing experiments may, perhaps, induce the author himself to think some such additions proper.

According to this hypothesis, we must suppose, that water consists of inflammable air united to dephlogisticated air; that nitrous air, vitriolic acid air, and the phosphoric acid, are also combinations of phlogisticated air, sulphur, and phosphorus, with dephlogisticated air; and that the two former, by a further addition of the same substance, are reduced to the common nitrous and vitriolic acids; that the metallic calces consist of the metals themselves united to the same substance, commonly, however, with a mixture of fixed air; that on exposing the calces of the perfect metals to a sufficient heat, all the dephlogisticated air is driven off, and the calces are restored to their metallic form: but as the calces of the imperfect metals are vitrified by heat, instead of recovering the metallic form, it should seem as if all the dephlogisticated air could not be driven off from them by heat alone. In like manner, according to this hypothesis, the rationale of the production of dephlogisticated air from red precipitate is, that during the solution of the quicksilver in the acid and the subsequent calcination, the acid is decompounded, and quits part of its dephlogisticated air to the quicksilver, whereby it comes over in the form of nitrous air, and leaves the quicksilver behind united to dephlogisticated air, which, by a further increase of heat, is driven off, while the quicksilver re-assumes its metallic form. In procuring dephlogisticated air from nitre, the acid is also decompounded; but with this difference, that it suffers some of its dephlogisticated air to escape, while it remains united to the alkali itself, in the form of phlogisticated nitrous acid. As to the production of dephlogisticated air from plants, it may be said, that vegetable substances consist chiefly of various combinations of three different bases, one of which, when united to dephlogisticated air, forms water, another fixed air, and the third phlogisticated air ; and that by means of vegetation each of these substances are decomposed, and yield their dephlogisticated air ; and that in burning they again acquire dephlogisticated air, and are restored to their pristine form.

It seems, therefore, from what has been said, as if the phænomena of nature might be explained very well on this principle, without the help of phlogiston ; and indeed, as adding dephlogisticated air to a body comes to the same thing as depriving it of its phlogiston and adding water to it, and as there are, perhaps, no bodies entirely destitute of water, and as I know no way by which phlogiston can be transferred from one body to another, without leaving it uncertain whether water is not at the same time transferred, it will be very difficult to determine by experiment which of these opinions is the truest ; but as the commonly received principle of phlogiston explains all phænomena, at least as well as Mr. Lavoisier's, I have adhered to that. There is one circumstance also, which though it may appear to many not to have much force, I own has some weight with me ; it is, that as plants seem to draw their nourishment almost intirely from water and fixed and phlogisticated air, and are restored back to those substances by burning, it seems reasonable to conclude, that notwithstanding their infinite variety they consist almost entirely of various combinations of water and fixed and phlogisticated air, united according to one of these opinions to phlogiston, and deprived according to the other of dephlogisticated air; so that, according to the latter opinion, the substance of a plant is less compounded than a mixture of those bodies into which it is resolved by burning; and it is more reasonable to look for great variety in the more compound than in the more simple substance.

Another thing which Mr. Lavoisier endeavours to prove is, that dephlogisticated air is the acidifying principle. From what has been explained it appears, that this is no more than saying, that acids lose their acidity by uniting to phlogiston, which with regard to the nitrous, vitriolic, phosphoric, and arsenical acids is certainly true. The same thing, I believe, may be said of the acid of sugar; and Mr. Lavoisier's experiment is a strong confirmation of Bergman's opinion, that none of the spirit of nitre enters into the composition of the acid, but that it only serves to deprive the sugar of part of its phlogiston. But as to the marine acid and acid of tartar, it does not appear that they are capable of losing their acidity by any union with phlogiston. It is to be remarked also, that the acids of sugar and tartar, and in all probability almost all the vegetable and animal acids, are by burning reduced to fixed and phlogisticated air, and water, and therefore contain more phlogiston, or less dephlogisticated air, than those three substances.

SECOND PAPER.

Philosophical Transactions, Vol. 75 (for 1785), pp. 372-384.

Read June 2, 1785.

IN a Paper, printed in the last volume of the Philosophical Transactions, in which I gave my reasons for thinking that the diminution produced in atmospheric air by phlogistication is not owing to the generation of fixed air, I said it seemed most likely, that the phlogistication of air by the electric spark was owing to the burning of some inflammable matter in the apparatus; and that the fixed air, supposed to be produced in that process, was only separated from that inflammable matter by the burning. At that time, having made no experiments on the subject myself, I was obliged to form my opinion from those already published; but I now find, that though I was right in supposing the phlogistication of the air does not proceed from phlogiston communicated to it by the electric spark, and that no part of the air is converted into fixed air; yet that the real cause of the diminution is very different from what I suspected, and depends upon the conversion of phlogisticated air into nitrous acid.

The apparatus used in making the experiments was as follows. The air through which the spark was intended to be passed, was confined in a glass tube M, bent to an angle, as in fig. 1. which, after being filled with quicksilver, was inverted into two glasses of the same fluid, as in the figure. The air to be tried was then introduced by means of a small tube, such as is used for ther-

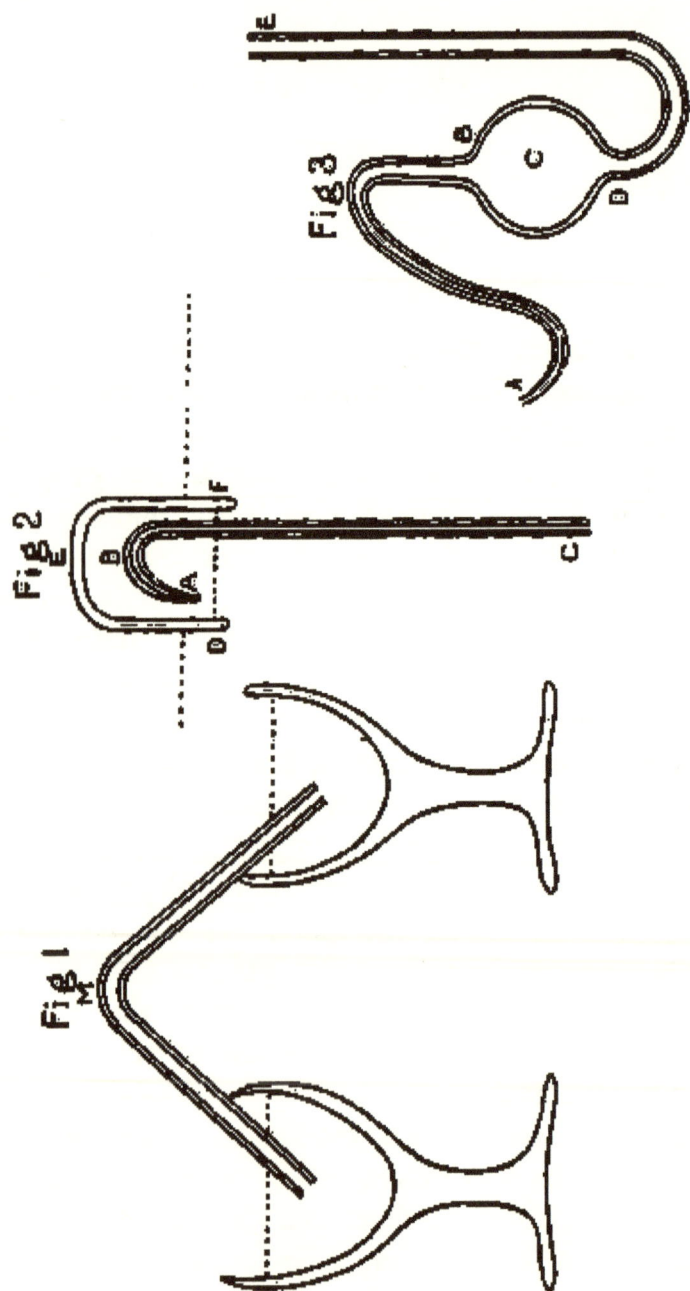

Fig 1

Fig 2

Fig 3

mometers, bent in the manner represented by ABC
(fig. 2.) the bent end of which, after being previously
filled with quicksilver, was introduced, as in the figure,
under the glass DEF, inverted into water, and filled with
the proper kind of air, the end C of the tube being kept
stopped by the finger; then, on removing the finger from
C, the quicksilver in the tube descended in the leg BC,
and its place was supplied with air from the glass DEF.
Having thus got the proper quantity of air into the tube
ABC, it was held with the end C uppermost, and stopped
with the finger; and the end A, made smaller for that
purpose, being introduced into one end of the bent tube
M (fig. 1.), the air, on removing the finger from C, was
forced into that tube by the pressure of the quicksilver
in the leg BC. By these means I was enabled to introduce the exact quantity I pleased of any kind of air into
the tube M; and, by the same means, I could let up
any quantity of soap-lees, or any other liquor which I
wanted to be in contact with the air.

In one case, however, in which I wanted to introduce
air into the tube many times in the same experiment, I
used the apparatus represented in fig. 3. consisting of a
tube AB of a small bore, a ball C, and a tube DE of a
larger bore. This apparatus was first filled with quicksilver; and then the ball C, and the tube AB, were filled
with air, by introducing the end A under a glass inverted
into water, which contained the proper kind of air, and
drawing out the quicksilver from the leg ED by a
syphon. After being thus furnished with air, the apparatus was weighed, and the end A introduced into one
end of the tube M, and kept there during the experiment; the way of forcing air out of this apparatus into
the tube being by thrusting down the tube ED a wooden
cylinder of such a size as almost to fill up the whole bore,
and by occasionally pouring quicksilver into the same

tube, to supply the place of that pushed into the ball C. After the experiment was finished, the apparatus was weighed again, which shewed exactly how much air had been forced into the tube M during the whole experiment; it being equal in bulk to a quantity of quicksilver, whose weight was equal to the increase of weight of the apparatus.

The bore of the tube M used in most of the following experiments, was about one-tenth of an inch; and the length of the column of air, occupying the upper part of the tube, was in general from $1\frac{1}{2}$ to $\frac{3}{4}$ of an inch.

It is scarcely necessary to inform any one used to electrical experiments, that in order to force an electrical spark through the tube, it was necessary, not to make a communication between the tube and the conductor, but to place an insulated ball at such a distance from the conductor as to receive a spark from it, and to make a communication between that ball and the quicksilver in one of the glasses, while the quicksilver in the other glass communicated with the ground.

I now proceed to the experiments.

When the electric spark was made to pass through common air, included between short columns of a solution of litmus, the solution acquired a red colour, and the air was diminished, conformably to what was observed by Dr. Priestley.

When lime-water was used instead of the solution of litmus, and the spark was continued till the air could be no further diminished, not the least cloud could be perceived in the lime-water; but the air was reduced to two-thirds of its original bulk; which is a greater diminution than it could have suffered by mere phlogistication, as that is very little more than one-fifth of the whole.

The experiment was next repeated with some impure

dephlogisticated air. The air was very much diminished, but without the least cloud being produced in the lime-water. Neither was any cloud produced when fixed air was let up to it; but on the further addition of a little caustic volatile alkali, a brown sediment was immediately perceived.

Hence we may conclude, that the lime-water was saturated by some acid formed during the operation; as in this case it is evident, that no earth could be precipitated by the fixed air alone, but that caustic volatile alkali, on being added, would absorb the fixed air, and thus becoming mild, would immediately precipitate the earth; whereas, if the earth in the lime-water had not been saturated with an acid, it would have been precipitated by the fixed air. As to the brown colour of the sediment, it most likely proceeded from some of the quicksilver having been dissolved.

It must be observed, that if any fixed air, as well as acid, had been generated in these two experiments with the lime-water, a cloud must have been at first perceived in it, though that cloud would afterwards disappear by the earth being re-dissolved by the acid; for till the acid produced was sufficient to dissolve the whole of the earth, some of the remainder would be precipitated by the fixed air; so that we may safely conclude, that no fixed air was generated in the operation.

When the air is confined by soap-lees, the diminution proceeds rather faster than when it is confined by lime-water; for which reason, as well as on account of their containing so much more alkaline matter in proportion to their bulk, soap-lees seemed better adapted for experiments designed to investigate the nature of this acid, than lime-water. I accordingly made some experiments to determine what degree of purity the air should be of, in order to be diminished most readily, and to the

greatest degree; and I found, that, when good dephlogisticated air was used, the diminution was but small;
when perfectly phlogisticated air was used, no sensible
diminution took place; but when five parts of pure
dephlogisticated air were mixed with three parts of common air, almost the whole of the air was made to
disappear.

It must be considered, that common air consists of
one part of dephlogisticated air, mixed with four of
phlogisticated; so that a mixture of five parts of pure
dephlogisticated air, and three of common air, is the
same thing as a mixture of seven parts of dephlogisticated
air with three of phlogisticated.

Having made these previous trials, I introduced into
the tube a little soap-lees, and then let up some dephlogisticated and common air, mixed in the above-mentioned proportions, which rising to the top of the tube
M, divided the soap-lees into its two legs. As fast as
the air was diminished by the electric spark, I continued
adding more of the same kind, till no further diminution
took place: after which a little pure dephlogisticated air,
and after that a little common air, were added, in order
to see whether the cessation of diminution was not
owing to some imperfection in the proportion of the two
kinds of air to each other; but without effect *. The
soap-lees being then poured out of the tube, and

---

* From what follows it appears, that the reason why the air
ceased to diminish was, that as the soap-lees were then become
neutralized, no alkali remained to absorb the acid formed by the
operation, and in consequence scarce any air was turned into acid.
The spark, however, was not continued long enough after the apparent cessation of diminution, to determine with certainty, whether
it was only that the diminution went on remarkably slower than
before, or that it was almost come to a stand, and could not have
been carried much further, though I had persisted in passing the
sparks.

separated from the quicksilver, seemed to be perfectly neutralized, as they did not at all discolour paper tinged with the juice of blue flowers. Being evaporated to dryness, they left a small quantity of salt, which was evidently nitre, as appeared by the manner in which paper, impregnated with a solution of it, burned.

For more satisfaction, I tried this experiment over again on a larger scale. About five times the former quantity of soap-lees were now let up into a tube of a larger bore ; and a mixture of dephlogisticated and common air, in the same proportions as before, being introduced by the apparatus represented in fig. 3. the spark was continued till no more air could be made to disappear. The liquor, when poured out of the tube, smelled evidently of phlogisticated nitrous acid, and being evaporated to dryness, yielded $1\frac{3}{10}$ gr. of salt, which is pretty exactly equal in weight to the nitre which that quantity of soap-lees would have afforded if saturated with nitrous acid. This salt was found, by the manner in which paper dipped into a solution of it burned, to be true nitre. It appeared, by the test of *terra ponderosa salita*, to contain not more vitriolic acid than the soap-lees themselves contained, which was excessively little ; and there is no reason to think that any other acid entered into it, except the nitrous.

A circumstance, however, occurred, which at first seemed to shew, that this salt contained some marine acid ; namely, an evident precipitation took place when a solution of silver was added to some of it dissolved in water ; though the soap-lees used in its formation were perfectly free from marine acid, and though, to prevent all danger of any precipitate being formed by an excess of alkali in it, some purified nitrous acid had been added to it, previous to the addition of the solution of silver. On consideration, however, I suspected, that this precipitation might arise from the nitrous acid in it being phlogisticated ; and therefore I tried whether nitre, much phlogisticated, would precipitate silver from its solution. For this purpose I exposed some nitre to the fire, in an earthen retort, till it had yielded a good deal of dephlogisticated air ; and then, having dissolved it in water, and added to it some well purified spirit of nitre till it was sensibly acid, in order to be certain that the alkali did not predominate, I dropped into it some solution of silver, which immediately made a very copious precipitate. This solution, however, being deprived of some of its phlogiston by evaporation to dryness, and exposure for a few weeks to the air, lost the property of precipitating silver from its solution ; a proof that this property depended only on its phlogistication, and not on its having absorbed sea-salt from the retort, or by any other means.

Hence it is certain, that nitre, when much phlogisticated, is capable of making a precipitate with a solution of silver ; and therefore there is no reason to think, that the precipitate, which our salt occasioned with a solution of silver, proceeded from any other cause than that of its being phlogisticated ; especially as it appeared by the smell, both on first taking it out of the tube, and on the addition of the spirit of nitre, previous to dropping in the solution of silver, that the acid in it was much phlogisticated. This property of phlogisticated nitre is worth the attention of chemists ; as otherwise they may sometimes be led into mistakes, in investigating the presence of marine acid by a solution of silver.

In the above-mentioned Paper I said, that when nitre is detonated with charcoal, the acid is converted into phlogisticated air ; that is, into a substance which, as far as I could perceive, possesses all the properties of the phlogisticated air of our atmosphere ; from which I concluded, that phlogisticated air is nothing else than nitrous acid united to phlogiston. According to this conclusion, phlogisticated air ought to be reduced to nitrous acid by being deprived of its phlogiston. But as dephlogisticated air is only water deprived of phlogiston, it is plain, that adding dephlogisticated air to a body, is equivalent to depriving it of phlogiston, and adding water to it ; and therefore, phlogisticated air ought also to be reduced to nitrous acid, by being made to unite to, or form a chemical combination with, dephlogisticated air; only the acid formed this way will be more dilute, than if the phlogisticated air was simply deprived of phlogiston.

This being premised, we may safely conclude, that in the present experiments the phlogisticated air was enabled, by means of the electrical spark, to unite to, or form a chemical combination with, the dephlogisticated air, and was thereby reduced to nitrous acid, which united to the soap-lees, and formed a solution of nitre ; for in these experiments those two airs actually disappeared, and nitrous acid was actually formed in their room ; and as, moreover, it has just been shewn, from other circumstances, that phlogisticated air must form nitrous acid, when combined with dephlogisticated air, the above-mentioned opinion seems to be sufficiently established. A further confirmation of it is, that, as far as I can perceive, no diminution of air is produced when the electric spark is passed either through pure dephlogisticated air, or through perfectly phlogisticated air ; which indicates the necessity of a combination of these two airs to produce the acid. Moreover, it was found in the last experiment, that the quantity of nitre procured was the same that the soap-lees would have produced if saturated with nitrous acid ; which shews, that the production of the nitre was not owing to any decomposition of the soap-lees.

It may be worth remarking, that whereas in the detonation of nitre with inflammable substances, the acid unites to phlogiston, and forms phlogisticated air, in these experiments the reverse of this process was carried on; namely, the phlogisticated air united to the dephlogisticated air, which is equivalent to being deprived of its phlogiston, and was reduced to nitrous acid.

In the above-mentioned Paper I also gave my reasons for thinking, that the small quantity of nitrous acid, produced by the explosion of dephlogisticated and inflammable air, proceeded from a portion of phlogisticated air mixed with the dephlogisticated, which I supposed was deprived of its phlogiston, and turned into nitrous acid, by the action of the dephlogisticated air on it, assisted by the heat of the explosion. This opinion, as must appear to every one, is confirmed in a remarkable manner by the foregoing experiments; as from them it is evident, that dephlogisticated air is able to deprive phlogisticated air of its phlogiston, and reduce it into acid, when assisted by the electric spark; and therefore it is not extraordinary that it should do so, when assisted by the heat of the explosion.

The soap-lees used in the foregoing experiments were made from salt of tartar, prepared without nitre; and were of such a strength as to yield one-tenth of their weight of nitre when saturated with nitrous acid. The dephlogisticated air also was prepared without nitre, that used in the first experiment with the soap-lees being procured from the black powder formed by the agitation of quicksilver mixed with lead *, and that used in the latter from turbith mineral. In the first experiment, the

---

* This air was as pure as any that can be procured by most processes. I propose giving an account of the experiment, in which it was prepared, in a future Paper.

quantity of soap-lees used was 35 measures, each of which was equal in bulk to one grain of quicksilver ; and that of the air absorbed was 416 such measures of phlogisticated air, and 914 of dephlogisticated. In the second experiment, 178 measures of soap-lees were used, and they absorbed 1920 of phlogisticated air, and 4860 of dephlogisticated. It must be observed, however, that in both experiments some air remained in the tube uncondensed, whose degree of purity I had no way of trying: so that the proportion of each species of air absorbed is not known with much exactness.

As far as the experiments hitherto published extend, we scarcely know more of the nature of the phlogisticated part of our atmosphere, than that it is not diminished by lime-water, caustic alkalies, or nitrous air; that it is unfit to support fire, or maintain life in animals; and that its specific gravity is not much less than that of common air : so that, though the nitrous acid, by being united to phlogiston, is converted into air possessed of these properties, and consequently, though it was reasonable to suppose, that part at least of the phlogisticated air of the atmosphere consists of this acid united to phlogiston, yet it might fairly be doubted whether the whole is of this kind, or whether there are not in reality many different substances confounded together by us under the name of phlogisticated air. I therefore made an experiment to determine, whether the whole of a given portion of the phlogisticated air of the atmosphere could be reduced to nitrous acid, or whether there was not a part of a different nature from the rest, which would refuse to undergo that change. The foregoing experiments indeed in some measure decided this point, as much the greatest part of the air let up into the tube lost its elasticity ; yet, as some remained unabsorbed, it did not appear for certain whether that was of the same nature as the rest

or not. For this purpose I diminished a similar mixture
of dephlogisticated and common air, in the same manner
as before, till it was reduced to a small part of its original
bulk. I then, in order to decompound as much as I
could of the phlogisticated air which remained in the
tube, added some dephlogisticated air to it, and continued the spark till no further diminution took place.
Having by these means condensed as much as I could
of the phlogisticated air, I let up some solution of liver
of sulphur to absorb the dephlogisticated air; after which
only a small bubble of air remained unabsorbed, which
certainly was not more than $\frac{1}{120}$ of the bulk of the phlogisticated air let up into the tube; so that if there is any
part of the phlogisticated air of our atmosphere which
differs from the rest, and cannot be reduced to nitrous
acid, we may safely conclude, that it is not more than
$\frac{1}{120}$ part of the whole.

The foregoing experiments shew, that the chief cause
of the diminution which common air, or a mixture of
common and dephlogisticated air, suffers by the electric
spark, is the conversion of the air into nitrous acid; but
yet it seemed not unlikely, that when any liquor, containing inflammable matter, was in contact with the air
in the tube, some of this matter might be burnt by the
spark, and thereby diminish the air, as I supposed in the
above-mentioned Paper to be the case. The best way
which occurred to me of discovering whether this hap-
pened or not, was to pass the spark through dephlogisticated air, included between different liquors: for
then, if the diminution proceeded solely from the conversion of air into nitrous acid, it is plain that, when the
dephlogisticated air was perfectly pure, no diminution
would take place; but when it contained any phlogisticated air, all this phlogisticated air, joined to as much of
the dephlogisticated air as must unite to it in order to

reduce it into acid, that is, two or three times its bulk, would disappear, and no more; so that the whole diminution could not exceed three or four times the bulk of the phlogisticated air: whereas, if the diminution proceeded from the burning of the inflammable matter, the purer the dephlogisticated air was, the greater and quicker would be the diminution.

The result of the experiments was, that when dephlogisticated air, containing only $\frac{1}{20}$ of its bulk of phlogisticated air (that being the purest air I then had), was confined between short columns of soap-lees, and the spark passed through it till no further diminution could be perceived, the air lost $\frac{143}{200}$ of its bulk; which is not a greater diminution than might very likely proceed from the first-mentioned cause; as the dephlogisticated air might easily be mixed with a little common air while introducing into the tube.

When the same dephlogisticated air was confined between columns of distilled water, the diminution was rather greater than before, and a white powder was formed on the surface of the quicksilver beneath; the reason of which, in all probability, was, that the acid produced in the operation corroded the quicksilver, and formed the white powder; and that the nitrous air, produced by that corrosion, united to the dephlogisticated air, and caused a greater diminution than would otherwise have taken place.

When a solution of litmus was used, instead of distilled water, the solution soon acquired a red colour, which grew paler and paler as the spark was continued, till at last it became quite colourless and transparent. The air was diminished by almost half, and I believe might have been still further diminished, had the spark been continued. When lime-water was let up into the tube, a cloud was formed, and the air was further diminished by about one-fifth. The remaining air was good dephlogisticated air. In this experiment, therefore, the litmus was, if not burnt, at least decompounded, so as to lose entirely its purple colour, and to yield fixed air ; so that, though soap-lees cannot be decompounded by this process, yet the solution of litmus can, and so very likely might the solutions of many other combustible substances. But there is nothing, in any of these experiments, which favours the opinion of the air being at all diminished by means of phlogiston communicated to it by the electric spark.

PRINTED AT THE DARIEN PRESS, EDINBURGH.